SOLAR POWER FOR BEGINNERS

A BEGINNER'S GUIDE TO INSTALL AND MAINTAIN SOLAR POWER

Copyright@2023

Carlos Alejandro

TABLE OF CONTENT

CHAPTER 1. INTRODUCTION

IMPORTANCE OF RENEWABLE ENERGY

The importance of renewable energy, such as solar power, cannot be overstated in today's world. By embracing renewable energy sources like solar power, we can mitigate climate change, protect the environment, foster economic growth, improve public health, and create a sustainable and equitable future for all. Some reasons why renewable energy is crucial:

1. Climate Change Mitigation: Renewable energy sources produce significantly fewer greenhouse gas emissions compared to fossil fuels. By reducing our reliance on fossil fuels and transitioning to renewables, we can mitigate climate change and limit the negative impacts of global warming.

2. Environmental Preservation: Traditional energy sources like coal and oil extraction have significant environmental consequences, including air and water pollution, habitat destruction, and ecosystem degradation. Renewable energy, on the other hand, has minimal environmental impact, helping to preserve ecosystems and biodiversity.

3. Energy Security and Independence:
Renewable energy diversifies our energy sources and reduces dependence on finite fossil fuel reserves, which are subject to geopolitical tensions and price volatility. Investing in renewables promotes energy security by utilizing local resources and reducing reliance on imported fuels.

4. Economic Opportunities: The renewable energy sector presents significant economic opportunities. It stimulates job creation across various industries, including manufacturing, installation, maintenance, and research and development. Renewable energy investments can revitalize local economies, attract investment, and foster innovation.

5. Public Health Benefits: Traditional energy sources are associated with air and water pollution, leading to adverse health effects such as respiratory diseases, cardiovascular issues, and premature deaths. Shifting to renewable energy reduces these pollutants, improving public health outcomes and reducing healthcare costs.

6. Sustainable Development: Renewable energy aligns with the principles of sustainable development by meeting present energy needs without compromising the ability of future generations to meet their own needs. It offers a long-term, sustainable energy solution that balances economic growth, social equity, and environmental stewardship.

7. Energy Access and Equity: Renewable energy provides opportunities for energy access and equity, particularly in underserved areas or developing countries. Solar power, for instance, can bring electricity to remote regions without access to centralized grids, empowering communities and enhancing quality of life.

8. Technological Innovation: Investing in renewable energy drives technological advancements. Research and development in solar power, wind energy, energy storage, and other renewables foster innovation, leading to more efficient and cost-effective technologies that benefit various sectors beyond energy.

9. Resilience and Adaptation: Renewable energy systems, especially distributed ones like solar panels on rooftops, enhance resilience to power outages and natural disasters. Localized energy generation and storage enable communities to maintain critical services and recover more quickly from disruptions.

10. Global Cooperation: The transition to renewable energy requires international collaboration. Governments, organizations, and individuals must work together to address climate change, promote sustainable development, and ensure a clean energy future for generations to come.

BENEFITS OF SOLAR POWER

Solar power offers numerous benefits that contribute to its growing popularity as a renewable energy source.

1. Renewable and Sustainable: Solar power relies on the energy of the sun, which is an abundant and inexhaustible source of energy. Unlike fossil fuels, solar power is renewable, ensuring a sustainable energy supply for future generations.

2. Reduction of Greenhouse Gas Emissions: Solar power generates electricity without producing greenhouse gas emissions, unlike fossil fuels that release carbon dioxide and other harmful pollutants. By using solar power, we can significantly reduce our carbon footprint and mitigate climate change.

3. Cost Savings: Installing solar panels can lead to substantial cost savings over time. Once the initial installation cost is recouped, solar power can provide free electricity, reducing or eliminating monthly utility bills. Additionally, excess electricity generated can be sold back to the grid through net metering, further reducing costs.

4. Energy Independence: Solar power allows individuals and communities to become more energy independent. By generating electricity on-site, solar power reduces dependence on centralized power grids and imported fossil fuels. This can enhance energy security and resilience, particularly during power outages or emergencies.

5. Long-Term Financial Investment: Installing solar panels is a long-term investment that can provide a reliable return on investment (ROI). Solar systems have a long lifespan and require minimal maintenance, resulting in long-term financial benefits and potential increased property value.

6. Job Creation and Economic Growth: The solar power industry creates jobs and stimulates economic growth. From manufacturing solar panels to installation and maintenance, solar power supports employment opportunities across the supply chain, contributing to local economies.

7. Scalability and Flexibility: Solar power systems are scalable, meaning they can be customized to fit different energy needs, whether it's for residential, commercial, or utility-scale applications. Solar panels can be installed on rooftops, open land, or integrated into various structures, offering flexibility in deployment.

8. Reduced Air Pollution and Improved Public Health: Solar power significantly reduces air pollution associated with traditional energy sources, improving air quality and public health. By displacing fossil fuel-based electricity generation, solar power helps reduce respiratory and cardiovascular diseases and related health costs.

9. Low Operational and Maintenance Costs:
Solar power systems have relatively low operational and maintenance costs compared to traditional power plants. Once installed, solar panels require minimal upkeep, with routine inspections and occasional cleaning being the primary maintenance tasks.

10. Technological Advancements and Innovation: The solar power industry continues to advance rapidly, driving technological innovation. Research and development efforts result in more efficient solar panels, energy storage solutions, and integration with smart grid technologies, improving overall performance and expanding the range of solar applications.

CHAPTER II. UNDERSTANDING SOLAR POWER

SOLAR ENERGY BASICS

Solar energy is derived from the radiant light and heat emitted by the sun. It is a renewable energy source that can be converted into usable electricity or used directly for heating and lighting purposes. Some key solar energy basics:

1. Solar Radiation: The sun emits energy in the form of electromagnetic radiation, including visible light, ultraviolet (UV) rays, and infrared (IR) radiation. Solar radiation is the fuel for solar energy systems.

2. Solar Panels (Photovoltaic Cells): Solar panels, also known as photovoltaic (PV) panels, are devices that convert sunlight into electricity through the photovoltaic effect. They are made up of multiple solar cells, typically composed of semiconductor materials, such as silicon.

3. Photovoltaic Effect: The photovoltaic effect is the process by which solar cells convert sunlight into electricity. When sunlight strikes the solar cells, it excites electrons in the semiconductor material, creating an electric current.

4. Solar Thermal Systems: Solar thermal systems use the heat from solar radiation to generate hot water, space heating, or to drive cooling systems. These systems typically use solar collectors, such as flat-plate or evacuated tube collectors, to absorb and transfer heat.

5. Solar Energy Applications: Solar energy can be utilized in various applications:

- **Solar Photovoltaic (PV) Systems:** PV systems convert sunlight directly into electricity and can power homes, buildings, or even entire communities.

- **Solar Water Heating Systems:** These systems use solar collectors to heat water for domestic or commercial use, such as showers, pools, and industrial processes.

- **Solar Space Heating and Cooling:** Solar energy can be used to heat or cool indoor spaces using solar thermal or PV technologies.

- **Solar-Powered Outdoor Lighting:** Solar panels can power outdoor lighting systems, such as streetlights, garden lights, and pathway lights.

- **Solar-Powered Ventilation:** Solar-powered ventilation systems use solar energy to power fans or vents, improving air circulation and reducing energy consumption.

6. Solar Energy Availability and Variability: The availability of solar energy depends on geographic location, time of year, and weather conditions. Areas with abundant sunlight receive more solar energy and are thus more suitable for solar energy systems. However, even regions with less sunlight can still harness solar energy effectively.

7. Net Metering: Net metering is a billing arrangement that allows solar energy system owners to receive credit for excess electricity they generate and feed back into the grid. It enables a two-way flow of electricity between the solar system and the grid, providing financial benefits to the system owner.

8. Solar Energy Storage: Solar energy can be stored for later use using energy storage technologies, such as batteries. Energy storage systems allow solar-generated electricity to be utilized during periods of low sunlight or at night, improving energy self-sufficiency and grid resiliency.

9. Environmental Benefits: Solar energy is a clean and environmentally friendly energy source. It does not produce greenhouse gas emissions during electricity generation, contributing to the reduction of air pollution and mitigating climate change.

10. Costs and Affordability: The cost of solar energy systems has significantly decreased over the years, making solar power more affordable and accessible. Government incentives, tax credits, and financing options further promote the adoption of solar energy systems. Understanding the basics of solar energy provides a foundation for exploring its applications, benefits, and potential for a sustainable and renewable energy future.

The sun as a source of energy

The sun is a massive, continuously shining star located at the center of our solar system. It serves as an abundant and virtually inexhaustible source of energy for our planet.

1. Nuclear Fusion: The sun generates energy through a process called nuclear fusion. In its core, hydrogen atoms fuse together to form helium, releasing an enormous amount of energy in the process. This fusion process is sustained by the sun's immense gravitational pressure and high temperatures.

2. Solar Radiation: The energy produced by the sun is emitted in the form of solar radiation. Solar radiation consists of various components, including visible light, ultraviolet (UV) rays, and infrared (IR) radiation. These forms of radiation travel through space and reach the Earth.

3. Solar Constant: The solar constant is the average amount of solar radiation that reaches the outer atmosphere of Earth. It is approximately 1361 watts per square meter (W/m^2). The solar constant represents the steady output of energy from the sun, providing a baseline for solar energy calculations.

4. Sun-Earth Distance: The distance between the sun and Earth is about 93 million miles (150 million kilometers). This distance is known as an astronomical unit (AU). Despite this vast distance, the sun's energy reaches Earth in about 8 minutes and 20 seconds, traveling at the speed of light.

5. Solar Energy Reaching Earth: Not all solar radiation reaches the Earth's surface directly. Various factors, such as atmospheric absorption, scattering, and reflection, affect the amount of solar energy that reaches the Earth's surface. However, even with these factors, the sun provides an abundant and substantial amount of energy.

6. Solar Energy Interactions: When solar radiation reaches the Earth's atmosphere and surface, it undergoes several interactions. Some of the energy is absorbed by the atmosphere, while a portion is reflected back into space. The energy that reaches the Earth's surface can be absorbed, reflected, or transmitted through different materials.

7. Harnessing Solar Energy: Humans have developed various technologies to harness the sun's energy for practical applications. Solar panels, also known as photovoltaic (PV) panels, convert sunlight directly into electricity using the photovoltaic effect. Solar thermal systems utilize solar collectors to capture and convert solar radiation into heat energy for water heating or space heating.

8. Renewable and Sustainable: The sun's energy is renewable and sustainable because it is continually produced through nuclear fusion. As long as the sun exists, it will continue to radiate energy, making it a virtually limitless source of power.

9. Solar Energy Potential: The potential of solar energy is vast. The amount of solar energy that reaches the Earth's surface in just one hour is enough to meet global energy demand for an entire year. By harnessing a fraction of this energy, we can reduce our dependence on fossil fuels and mitigate climate change.

10. Solar Energy and Beyond Earth: Solar energy is not limited to Earth. It powers the other planets in our solar system, and scientists have also explored the potential of utilizing solar energy in space missions and future human settlements on the moon or other celestial bodies. Understanding the sun as a source of energy helps us appreciate the immense power it provides and its potential for clean, renewable energy solutions. By harnessing solar energy, we can reduce our environmental impact, transition to sustainable energy sources, and create a more resilient and sustainable future.

HOW SOLAR PANELS WORK

Solar panels, also known as photovoltaic (PV) panels, convert sunlight directly into electricity through a process called the photovoltaic effect.

1. Solar Cells: Solar panels are made up of individual solar cells, which are typically composed of semiconductor materials, such as silicon. Each solar cell consists of two layers: a positive layer (p-type) and a negative layer (n-type).

2. Photons Absorption: When sunlight (which is composed of photons) strikes the surface of a solar panel, the photons are absorbed by the semiconductor material. This absorption of photons provides energy to the atoms in the material, causing electrons in the atoms to be released from their atoms' outer shells.

3. Electron Movement: The released electrons are then forced to move toward the positive layer of the solar cell due to the internal electric field created by the junction between the p-type and n-type layers. This movement of electrons creates a flow of electric current.

4. Electric Current Generation: The movement of the electrons creates a direct current (DC) flow within the solar cell. This DC current can then be harnessed for various applications, including charging batteries, powering electronic devices, or supplying electricity to a power grid.

5. Wiring and Circuitry: Multiple solar cells are interconnected within a solar panel to increase the overall voltage and current output. The cells are typically wired in a series or parallel configuration to achieve the desired electrical characteristics.

6. Inverter Conversion: Since most electrical devices and power grids operate on alternating current (AC), the DC electricity generated by the solar panels needs to be converted into AC. This is done using an inverter, which converts the DC electricity into AC electricity compatible with standard electrical systems.

7. Power Distribution: The AC electricity produced by the solar panels can be used to power electrical devices directly within a building or home. Excess electricity can also be fed back into the power grid, often through a process called net metering, where the utility company credits the solar panel owner for the electricity they generate.

8. System Monitoring: Solar panel systems often include monitoring systems that track the energy production and performance of the panels. This allows owners to monitor their energy generation and assess the system's efficiency.

It's important to note that the efficiency of solar panels can be influenced by factors such as the angle and orientation of the panels, shading, temperature, and the quality of the solar cells themselves. Advances in solar cell technology continue to improve the efficiency and affordability of solar panels, making them an increasingly popular and viable renewable energy option.

Photovoltaic (PV) vs. solar thermal systems

Photovoltaic (PV) systems and solar thermal systems are two distinct technologies used to harness solar energy. While both technologies utilize the sun's energy, they operate differently and serve different purposes. This is a comparison between PV and solar thermal systems:

1. Principle of Operation:

PV Systems: PV systems convert sunlight directly into electricity using the photovoltaic effect. Solar panels, consisting of multiple PV cells made of semiconductor materials, absorb photons from sunlight, causing the release of electrons and generating a flow of electric current.

Solar Thermal Systems: Solar thermal systems capture the sun's heat to generate thermal energy. Solar collectors, typically mounted on rooftops or in open areas, absorb solar radiation and transfer the heat to a fluid (such as water or a heat transfer fluid) flowing through the collectors. The heated fluid can then be used for various applications, such as water heating or space heating.

2. Energy Output:

PV Systems: PV systems generate electricity that can power electrical devices, charge batteries, or be fed into the electrical grid. The output is typically in the form of direct current (DC), which can be converted into alternating current (AC) using an inverter.

Solar Thermal Systems: Solar thermal systems produce heat energy that can be used for water heating, space heating, or industrial processes. The output is in the form of thermal energy, which can be used directly or stored for later use.

3. Applications:

PV Systems: PV systems are commonly used to generate electricity for residential, commercial, and utility-scale applications. They can power homes, businesses, schools, and even entire communities. PV systems can be installed on rooftops, ground-mounted in open areas, or integrated into building materials like solar tiles.

Solar Thermal Systems: Solar thermal systems are primarily used for water heating, space heating, and process heating. They are suitable for applications such as heating swimming pools, providing hot water for domestic use, or supplying heat to industrial processes.

4. Efficiency:

PV Systems: PV systems have made significant advancements in efficiency, with modern solar panels achieving high conversion rates. However, the efficiency of PV systems can be affected by factors like temperature, shading, and the quality of the solar cells.

Solar Thermal Systems: Solar thermal systems can have high overall system efficiency for converting solar radiation into usable heat. However, the efficiency can vary depending on factors like collector design, heat transfer fluid, and insulation.

5. Cost and Complexity:

PV Systems: PV systems have become increasingly affordable over the years, with declining costs of solar panels and associated components. They are generally simpler to install and maintain compared to solar thermal systems.

Solar Thermal Systems: Solar thermal systems may involve more complex installations, including plumbing and heat transfer systems. The equipment and installation costs can vary depending on the specific application and system size.

It's important to note that PV and solar thermal systems are not mutually exclusive, and they can be used together in certain applications. For example, a building may utilize PV panels to generate electricity and solar thermal collectors for water heating. Ultimately, the choice between PV and solar thermal systems depends on the specific energy needs and applications, as well as factors such as available space, cost considerations, and regional climate conditions.

COMPONENTS OF A SOLAR POWER SYSTEM

SOLAR PANELS

Solar panels, also known as photovoltaic (PV) panels, are devices that convert sunlight into electricity using the photovoltaic effect

1. Structure: Solar panels are typically rectangular in shape and consist of multiple interconnected solar cells. Each solar cell is made of semiconductor materials, such as silicon, that can absorb photons from sunlight.

2. Photovoltaic Effect: Solar panels harness the photovoltaic effect, which is the process of converting light (photons) into electricity. When sunlight strikes the surface of the solar panel, the photons transfer their energy to the electrons in the semiconductor material, causing them to be released from their atoms. This creates an electric current.

3. Solar Cell Construction: Solar cells are constructed with layers of different materials to facilitate the photovoltaic effect. The most common type of solar cell is made of a silicon wafer. The top layer is usually treated to have a negative charge (n-type), while the bottom layer is treated to have a positive charge (p-type).

4. Electrical Connections: Solar cells within a panel are connected in series or parallel configurations to achieve the desired voltage and current output. The cells are typically connected using metal conductors that allow the flow of electricity.

5. Encapsulation: To protect the solar cells from environmental factors and provide durability, they are encapsulated in a protective layer, typically made of tempered glass, which is transparent to sunlight. This layer also helps maximize light transmission to the solar cells.

6. Junction Box: A junction box is usually present on the backside of the solar panel. It contains diodes and other components that facilitate the electrical connection between the solar panel and the external electrical system, such as batteries or the power grid.

7. Efficiency: Solar panel efficiency refers to the ability of the panel to convert sunlight into electricity. Efficiency is influenced by factors such as the quality of the solar cells, the materials used, and manufacturing processes. Higher efficiency panels can produce more electricity from the same amount of sunlight.

8. Varieties of Solar Panels: There are diverse types of solar panels available:

- Monocrystalline Panels: Made from a single crystal structure, monocrystalline panels have high efficiency and a uniform black appearance.

- Polycrystalline Panels: Made from multiple crystal fragments, polycrystalline panels are slightly less efficient but often more cost-effective.

- Thin-Film Panels: Thin-film panels use a thin semiconductor layer deposited on a substrate, making them flexible and lightweight.

9. Installation: Solar panels are typically installed on rooftops or mounted on structures such as ground-mounted racks or carports. The optimal placement of solar panels takes into account factors like the orientation, tilt angle, shading, and available sunlight.

10. Applications: Solar panels are widely used for generating electricity in residential, commercial, and utility-scale applications. They can power homes, businesses, schools, and provide electricity to remote areas. Solar panels can also be integrated into various structures and devices like solar-powered streetlights, vehicles, and portable chargers.

Solar panels offer a clean and renewable energy solution, reducing reliance on fossil fuels and mitigating climate change. Continued advancements in technology are improving the efficiency and affordability of solar panels, making them an increasingly popular choice for sustainable energy generation.

INVERTER

An inverter is an essential component in a solar power system that converts the direct current (DC) electricity produced by solar panels into alternating current (AC) electricity. This is an overview of inverters and their role in solar power systems:

1. Function: The main function of an inverter is to convert the DC electricity generated by solar panels into AC electricity that is compatible with standard electrical systems. AC electricity is the type of electricity used in homes, businesses, and the power grid.

2. Conversion Process: The inverter uses electronic components and circuitry to convert the DC voltage and current from the solar panels into AC voltage and current. This conversion allows the electricity to be used for various applications or fed back into the grid.

3. Waveform: The output waveform of an inverter is typically a sine wave, which closely resembles the waveform of the electricity supplied by the utility grid. This ensures compatibility with a wide range of electrical devices and systems.

4. Maximum Power Point Tracking (MPPT): Many inverters are equipped with MPPT technology, which optimizes the performance of the solar panels by continuously tracking and adjusting the operating point to maximize power output. MPPT helps extract the maximum power from the solar panels, even under varying conditions like shading or temperature changes.

5. Grid-Tied Inverters: Grid-tied inverters are commonly used in solar power systems connected to the electrical grid. They synchronize the AC electricity produced by the solar panels with the grid's voltage and frequency. Grid-tied inverters also allow for net metering, where excess electricity generated by the solar system can be fed back into the grid, offsetting the owner's electricity consumption and potentially earning credits.

6. Off-Grid Inverters: Off-grid inverters are used in standalone solar systems that are not connected to the grid. They convert DC electricity from the solar panels into AC electricity to power electrical loads in off-grid applications, such as remote cabins, boats, or RVs. Off-grid inverters often incorporate battery charging capabilities to store excess energy for use during periods of low sunlight.

7. Hybrid Inverters: Hybrid inverters are designed for hybrid solar systems that combine solar panels with energy storage systems, such as batteries. These inverters manage the flow of electricity between the solar panels, batteries, and the electrical loads, allowing for self-consumption of solar energy and backup power during grid outages.

8. Monitoring and Control: Many inverters come with built-in monitoring and control functionalities. They provide real-time data on energy production, system performance, and enable remote monitoring and troubleshooting. Monitoring systems help users track their energy generation, identify issues, and optimize system performance.

9. Efficiency: Inverter efficiency refers to the percentage of DC power converted into usable AC power. Higher efficiency inverters minimize power losses during the conversion process, ensuring that more of the solar energy is effectively utilized.

10. Safety Features: Inverters often include safety features like ground fault protection, overvoltage protection, and temperature monitoring to ensure the safe operation of the solar power system.

BATTERIES (IF APPLICABLE)

Batteries play a vital role in solar power systems by storing excess electricity generated by solar panels for later use. Some overview of batteries in the context of solar power:

1. Energy Storage: Solar panels generate electricity when exposed to sunlight, but the amount of electricity produced can vary throughout the day and across different seasons. Batteries allow for the storage of surplus energy during periods of high solar generation for use during periods of low or no solar generation, such as at night or during cloudy days.

2. Types of Batteries: Different types of batteries are used in solar power systems, each with its own characteristics, advantages, and limitations. Common types include:

- **Lead-Acid Batteries:** These batteries are the most traditional and widely used type. They are relatively affordable and have a long-established track record. However, they are relatively heavy, require regular maintenance, and have a limited depth of discharge.

- **Lithium-Ion Batteries:** Lithium-ion batteries have gained popularity in recent years due to their high energy density, longer lifespan, lighter weight, and lower maintenance requirements. They offer a deeper depth of discharge and higher efficiency compared to lead-acid batteries.

- **Saltwater Batteries:** Saltwater batteries are a newer technology that utilizes a water-based electrolyte solution instead of traditional chemicals. They offer improved safety and environmental friendliness, but their efficiency and energy density may be lower compared to lithium-ion batteries.

3. Battery Capacity: The capacity of a battery refers to its ability to store energy, measured in kilowatt-hours (kWh). It determines the amount of electricity that can be stored and used when solar generation is low. The battery capacity should be chosen based on the energy consumption needs during non-generating periods, desired backup duration, and system design considerations.

4. Depth of Discharge (DoD): The depth of discharge refers to the percentage of a battery's capacity that can be safely discharged without damaging the battery. Batteries have recommended DoD limits to maintain their performance and lifespan. For example, a battery with a 50% DoD limit means it should not be discharged beyond 50% of its capacity to prolong its life.

5. Battery Management System (BMS): Many advanced battery systems incorporate a Battery Management System. The BMS monitors and controls the battery's charging and discharging processes, ensures optimal performance, prevents overcharging or over-discharging, and protects the battery from damage.

6. Sizing and Integration: Proper sizing and integration of batteries into a solar power system require careful consideration of energy consumption patterns, solar generation profile, desired backup capacity, and system design. Sizing calculations involve determining the number and capacity of batteries required to meet the energy storage needs of the system.

7. Off-Grid Systems vs. Grid-Tied Systems: Batteries are commonly used in off-grid solar systems where there is no access to the utility grid. They provide energy independence and allow for round-the-clock power supply. In grid-tied systems, batteries can be used for energy storage and backup power during grid outages, and they can also enable self-consumption of solar energy by storing excess electricity for later use.

8. Maintenance and Lifespan: Different battery chemistries have varying maintenance requirements and lifespans. Regular maintenance may include monitoring the state of charge, ensuring proper ventilation, and periodic inspections. Battery lifespan depends on factors such as usage patterns, temperature, charging and discharging rates, and the specific battery chemistry.

MOUNTING SYSTEMS AND WIRING

Mounting systems and wiring are crucial components of a solar power system that facilitate the installation, stability, and proper functioning of solar panels. Here's an overview of mounting systems and wiring in the context of solar installations:

Mounting Systems:

1. Roof Mounting: The most common method of

installing solar panels is roof mounting. Roof mounts are designed to securely attach solar panels to various types of roofs, such as asphalt shingle, metal, tile, or flat roofs. Different mounting options, including flush mounts, tilt mounts, or ballasted systems, may be used depending on the roof type, orientation, and local regulations.

2. Ground Mounting: Ground-mounted solar systems are installed on the ground or on structures like poles or racks. Ground mounts are suitable for situations where roof space is limited or when it is desirable to optimize panel tilt angle and orientation for maximum solar exposure.

3. Tracking Systems: Some solar installations use tracking systems that allow solar panels to follow the movement of the sun throughout the day, maximizing energy production. Tracking systems can be either single-axis (following the sun's path from east to west) or dual-axis (tracking both the sun's path and its elevation).

4. Racking and Supports: Racking systems provide the framework for mounting and positioning solar panels. They are typically made of metal and consist of rails, brackets, and clamps that securely hold the panels in place. Racking systems must be engineered to withstand wind, snow loads, and other environmental factors specific to the installation location.

5. Electrical Grounding: Proper grounding of the mounting system is essential for electrical safety. The mounting system should be bonded to the grounding system of the building or structure to provide a path for electrical fault currents and to protect against electrical hazards.

WIRING AND ELECTRICAL CONNECTIONS:

1. DC Wiring: Solar panels generate DC electricity, which needs to be routed from the panels to the inverter. DC wiring is typically made of high-quality, weather-resistant, and UV-stabilized cables designed to handle the current and voltage produced by the solar panels. Proper sizing and installation of DC wiring are crucial to minimize voltage drop and optimize system performance.

2. Combiner Boxes: In larger solar installations, combiner boxes are used to bring together the DC outputs of multiple strings of solar panels into a single connection point. Combiner boxes contain fuses or circuit breakers for overcurrent protection and can include other components like surge protectors and monitoring devices.

3. Inverter Connection: The DC wiring from the solar panels is connected to the input terminals of the inverter. The connection should be made following the manufacturer's specifications and electrical codes. The inverter then converts the DC electricity into AC electricity for use in the building or for feeding into the electrical grid.

4. AC Wiring: The AC electricity produced by the inverter is typically connected to the building's electrical distribution system. AC wiring follows standard electrical practices, including the use of appropriate wire sizes, conduit, circuit breakers, and grounding, as required by local electrical codes.

5. Electrical Protection: Proper electrical protection measures, such as overcurrent protection devices (breakers or fuses), surge protection devices, and disconnect switches, should be installed at appropriate points in the wiring system to ensure safety and compliance with electrical codes.

MONITORING AND CONTROL DEVICES

Monitoring and control devices are essential components of a solar power system that enable users to monitor, analyze, and manage the performance and operation of the system. This is an overview of monitoring and control devices commonly used in solar installations:

1. Inverter Monitoring: Many inverters come with built-in monitoring capabilities, allowing users to access real-time data on energy production, system performance, and other important parameters. Monitoring features may include information on power output, energy yield, voltage, current, and system status. Some inverters have integrated displays, while others can be monitored remotely through computer software or mobile apps.

2. Energy Monitoring Systems: Energy monitoring systems provide more comprehensive monitoring of the solar power system's performance. These systems can track not only the solar energy production but also the energy consumption of the building or facility. Energy monitoring devices can provide detailed information on energy usage patterns, peak demand, and overall system efficiency. They help users identify opportunities for energy optimization and cost savings.

3. Data Loggers and Communication Devices:
Data loggers are devices that collect and store data from various sensors and meters in the solar power system. They can capture data on energy production, energy consumption, weather conditions, and other relevant parameters. Communication devices, such as wireless or wired network interfaces, enable the transfer of data from the monitoring devices to a central monitoring system or cloud-based platform for analysis and reporting.

4. Monitoring Platforms and Software: There are dedicated software platforms and cloud-based solutions available for monitoring and analyzing solar power systems. These platforms aggregate and display the collected data in a user-friendly interface, allowing users to visualize and analyze the system's performance over time. They may provide features like customizable dashboards, data logging, reporting, and alert notifications for system faults or performance issues.

5. Remote Monitoring and Troubleshooting:

Remote monitoring capabilities allow users or solar system operators to access and monitor the system's performance from anywhere with an internet connection. Remote monitoring enables real-time troubleshooting, diagnostics, and performance analysis, reducing the need for physical inspections and maintenance visits. It also facilitates early detection of system faults or underperformance, ensuring timely intervention and maximizing system uptime.

6. Fault Detection and Alert Systems:

Monitoring devices and software can incorporate fault detection algorithms and alert systems. These features can automatically detect anomalies, such as low energy production, inverter malfunctions, or communication issues, and generate alerts to notify the system owner, installer, or maintenance personnel. Early fault detection helps minimize downtime and allows for prompt maintenance and repairs.

7. Performance Analysis and Reporting:

Monitoring systems often provide tools for analyzing historical data and generating reports on system performance. These tools can help evaluate the system's energy production, efficiency, and overall return on investment. Performance analysis and reporting aid in identifying trends, optimizing system performance, and providing documentation for warranty claims or financial incentives. Monitoring and control devices are valuable tools for solar system owners, installers, and maintenance teams to ensure optimal performance, troubleshoot issues, and maximize energy production. They provide insights into system performance, energy usage patterns, and enable proactive management to improve efficiency and reliability. Choosing monitoring devices and software that are compatible with the solar power system components and meet the specific monitoring requirements is important for effective system monitoring and control.

CHAPTER III. EVALUATING YOUR SOLAR POTENTIAL

ASSESSING SITE SUITABILITY

Assessing site suitability is a crucial step in the planning and design of a solar power system. These are some key factors to consider when evaluating the suitability of a site for solar installations:

1. Solar Resource: The solar resource of a site refers to the amount of sunlight it receives throughout the year. A site with ample sunlight exposure is essential for efficient energy production. Factors to consider include the site's latitude, average daily solar irradiation, and any potential shading from nearby buildings, trees, or obstacles.

2. Orientation and Tilt: The orientation and tilt angle of the solar panels affect their exposure to sunlight. The optimal orientation for solar panels in the Northern Hemisphere is typically facing south to maximize solar energy capture. The tilt angle should be adjusted based on the site's latitude to optimize energy generation throughout the year. East- and west-facing orientations can also be considered depending on specific objectives and energy consumption patterns.

3. Roof and Ground Space: The available roof or ground space is a critical factor in determining the size and capacity of the solar power system. Roof suitability includes factors such as the area, structural integrity, and orientation of the roof. For ground-mounted systems, the available land should have sufficient unobstructed space for the solar panels and meet any local zoning or setback requirements.

4. Shading: Shading from nearby buildings, trees, or other obstructions can significantly impact the performance of solar panels. It is important to assess potential shading throughout the day and across different seasons. Shade analysis tools or professional solar installers can help evaluate shading impacts and determine the best panel placement to minimize shading losses.

5. Grid Connectivity: If the intention is to connect the solar power system to the electrical grid, the proximity and accessibility to the grid connection point should be considered. Evaluating the existing electrical infrastructure and determining the feasibility and cost of grid interconnection is important for determining the project's viability.

6. Environmental Factors: Site assessments should also consider any environmental factors that could impact the solar power system, such as extreme weather conditions, wind patterns, snow accumulation, or the presence of corrosive elements. These factors can influence the selection of appropriate solar panel mounting systems, system design, and maintenance requirements.

7. Regulatory and Permitting Considerations: It is important to be aware of local regulations, zoning requirements, and permit processes that may affect solar installations. Familiarize yourself with any restrictions or requirements related to setbacks, height limitations, aesthetic considerations, or other local codes that may impact the design and installation of the system.

8. Financial Considerations: Assessing the financial viability of a solar power system includes evaluating the available incentives, such as tax credits, rebates, or feed-in tariffs, that can help offset the installation costs and provide a favorable return on investment. Additionally, conducting a financial analysis to estimate the system's potential energy savings and payback period is crucial in determining the financial feasibility of the project. Site assessments should be conducted by qualified solar professionals or engineers who can evaluate these factors and provide a detailed analysis of the site's suitability for solar installations. Site visits, solar modeling software, and tools like solar path finders can aid in gathering accurate data and performing thorough evaluations.

SUN EXPOSURE AND SHADING ANALYSIS

Sun exposure and shading analysis are important aspects of assessing the suitability of a site for solar installations. These are an overview of how sun exposure and shading analysis are conducted:

1. Sun Path Analysis: Sun path analysis involves studying the path of the sun across the sky throughout the year at a specific location. This analysis helps determine the solar azimuth (the angle between true south and the position of the sun) and solar altitude (the height of the sun in the sky) at different times of the day and seasons.

2. Sun Mapping Tools: Sun mapping tools, such as solar path finders or online solar calculators, can provide accurate data on sun angles and shading at a specific site. These tools use GPS coordinates, time, and date to generate a sun path diagram or a

3D representation of how sunlight will interact with the site's surroundings.

3. Shade Analysis: Shade analysis involves identifying and quantifying the shading obstacles that can affect solar panels. It is crucial to assess shading throughout the day and across different seasons, as the sun's position changes. Shade analysis can be conducted using various methods, including:

- **Physical Inspection:** A physical inspection of the site is conducted to identify potential shading sources, such as nearby buildings, trees, or structures. The height, distance, and orientation of shading objects are assessed to determine their impact on the solar panels.

- **Shade Analysis Tools:** Shade analysis tools use advanced software to model shading patterns based on the site's topography, nearby objects, and sun path data. These tools can simulate shading effects and generate reports or visual representations of

shading intensity and duration at specific
times.

- **Drone or Aerial Surveys:** Drone or aerial
  surveys can provide a bird's-eye view of the
  site, allowing for a comprehensive
  assessment of shading obstacles. High-
  resolution imagery can be used to identify
  potential shading sources and analyze their
  impact on the solar installation.

4. Shading Mitigation Strategies: If shading
analysis reveals significant shading impacts,
various strategies can be employed to mitigate or
minimize the effects, such as:

- **Tree Trimming or Removal:** Pruning or
  removing trees that cast shadows on the
  solar panels can help reduce shading effects.

- **Panel Placement and Tilt:** Adjusting the
  tilt angle or orientation of the solar panels
  can optimize sunlight exposure and
  minimize shading losses. Tilting the panels

to increase the height off the ground can also reduce shading from nearby objects.

- **Micro-Inverters or Power Optimizers:** Micro-inverters or power optimizers can be used to mitigate the impact of shading on the overall system performance. These devices allow each solar panel to operate independently, reducing the performance loss caused by shading on individual panels.

- **Solar Tracking Systems:** In some cases, solar tracking systems that follow the sun's movement throughout the day can be employed to minimize the impact of shading. Tracking systems ensure that panels are always positioned optimally to receive maximum sunlight.

ROOF ORIENTATION AND TILT

Roof orientation and tilt are important considerations when designing a solar power system. An overview of how roof orientation and tilt impact solar energy production:

ROOF ORIENTATION:

The optimal roof orientation for solar panels in the Northern Hemisphere is typically facing south (true south) to maximize solar energy capture. South-facing panels receive the most sunlight throughout the day, as the sun's path is predominantly from the south. However, east and west orientations can also be viable depending on specific objectives and energy consumption patterns. East-facing panels receive more sunlight in the morning, which can be beneficial if energy consumption is higher during the morning hours or if there are shading concerns in the afternoon. West-facing panels, on the other hand, receive more sunlight in the afternoon, which can be advantageous for those with higher energy usage in the afternoon or shading issues in the morning.

If south, east, or west orientations are not feasible, adjustments can be made to optimize energy production. For instance, panels can be installed at an azimuth angle slightly east or west of true south to capture more sunlight during specific periods of the day.

ROOF TILT:

The tilt angle of solar panels affects their exposure to sunlight and energy production. The ideal tilt angle depends on the latitude of the installation site and the seasonal variations in the sun's path. To determine the optimal tilt angle, one common approach is to set it equal to the site's latitude. This allows the panels to capture the most sunlight throughout the year. However, in some cases, adjusting the tilt angle based on seasonal variations may be more advantageous. During the winter months, when the sun is lower in the sky, increasing the tilt angle can enhance energy production by maximizing the capture of low-angle sunlight. Conversely, reducing the tilt angle during the summer months, when the sun is higher in the sky, can help prevent potential overproduction and maximize self-consumption of solar energy.

OTHER FACTORS INFLUENCING SOLAR POTENTIAL

In addition to roof orientation and tilt, several other factors can influence the solar potential and energy production of a solar power system. These factors include:

1. Solar Irradiance: Solar irradiance refers to the amount of solar energy that reaches a given area over a specific period. Regions with higher solar irradiance levels receive more sunlight and have greater solar potential. Factors such as latitude, local climate, cloud cover, and atmospheric conditions affect solar irradiance. Solar resource maps and databases provide information on average solar irradiance levels for different regions, aiding in assessing solar potential.

2. Climate and Weather Patterns: Local climate and weather patterns play a role in solar energy production. Areas with clear skies and a high number of sunny days throughout the year generally have greater solar potential. Cloud cover, humidity, and pollution levels can impact the amount of sunlight reaching the solar panels. It is important to consider historical weather data and seasonal variations in sunlight availability when assessing solar potential.

3. Temperature: Solar panels are affected by temperature, and high temperatures can reduce their efficiency. While solar panels generate electricity from sunlight, they actually perform better at cooler temperatures. The performance characteristics of solar panels under different temperature conditions should be taken into account when evaluating solar potential and system design.

4. Altitude and Elevation: Higher altitudes generally receive more solar radiation due to reduced atmospheric thickness and less cloud cover. However, it's important to consider other factors, such as temperature variations, wind patterns, and installation challenges at higher elevations.

5. Geographic Location and Obstructions: The geographic location of a site can influence solar potential. For example, sites in valleys or areas surrounded by tall buildings or mountains may experience shading during certain times of the day, reducing energy production. Assessing potential obstructions and their impact on sunlight availability is crucial in determining solar potential.

6. Roof Size and Available Space: The size of the roof or available space for solar panel installation affects the system's capacity and energy production. Larger roof areas or more extensive open spaces allow for the installation of a greater number of solar panels, resulting in higher energy output.

7. System Efficiency and Technology: The efficiency of solar panels and the overall system impacts energy production. More efficient solar panels can convert a higher percentage of sunlight into electricity, increasing overall system performance. Additionally, the choice of solar cell technology, such as monocrystalline, polycrystalline, or thin-film, can influence the system's energy production and suitability for specific site conditions.

8. Orientation of Nearby Structures: Nearby structures, such as buildings or trees, can cast shadows on the solar panels and affect energy production. Assessing the height, distance, and orientation of surrounding structures is crucial in determining potential shading impacts and optimizing panel placement.

9. Regulatory and Financial Incentives: The availability of regulatory incentives, such as feed-in tariffs, tax credits, rebates, or net metering programs, can impact the financial viability and attractiveness of a solar power system. Understanding and evaluating these incentives is important when considering the solar potential of a site.

CALCULATING ENERGY REQUIREMENTS

Analyzing electricity consumption

Analyzing electricity consumption is an important step in designing a solar power system and understanding its potential impact. Here are some key aspects to consider when analyzing electricity consumption:

1. Historical Electricity Bills: Reviewing historical electricity bills is a good starting point to understand the patterns of electricity consumption. Examine the monthly or annual usage data to identify trends, peak demand periods, and any seasonal variations. This information provides insights into the average and maximum electricity consumption levels, helping determine the system size needed to offset or reduce electricity usage.

2. Load Profiling: Load profiling involves monitoring and analyzing electricity usage patterns in more detail. It entails gathering data on a more granular level, such as hourly or daily usage, to identify specific periods of high or low energy demand. Load profiling can be done using smart meters, energy monitoring devices, or by manually recording electricity usage at regular intervals. This analysis helps identify opportunities for optimizing energy consumption, such as shifting high-demand activities to times when solar production is at its peak.

3. Energy Consumption Breakdown: Understanding how energy is used within a property is essential for system design and sizing. Analyze the breakdown of energy consumption by different appliances, equipment, or areas of the property. This information helps identify the major energy consumers and potential areas for energy efficiency improvements. For example, heating, ventilation, and air conditioning (HVAC) systems, lighting, and appliances often account for a significant portion of electricity usage.

4. Time-of-Use (TOU) Rates: Many utility companies offer time-of-use rates, where electricity costs vary depending on the time of day. Analyzing TOU rate structures and matching them with consumption patterns can help identify peak demand periods and strategize solar energy usage to maximize cost savings. Solar power systems can be designed to produce electricity during these peak demand periods, reducing reliance on grid electricity and optimizing financial benefits.

5. Energy Efficiency Measures: Before installing a solar power system, it is beneficial to implement energy efficiency measures to reduce overall electricity consumption. Conduct an energy audit or assessment to identify opportunities for energy savings, such as upgrading to energy-efficient appliances, improving insulation, using LED lighting, or implementing smart home technologies. Reducing energy consumption upfront can lead to a smaller, more cost-effective solar system size.

6. Future Energy Needs: Consider future changes in energy consumption patterns when analyzing electricity usage. Anticipate any planned changes in lifestyle, occupancy, or business operations that may affect electricity demand. For example, if you're planning to purchase electric vehicles or expand your business, factor in the additional electricity requirements in the analysis to ensure the solar system is adequately sized.

Estimating system size and capacity

Estimating the system size and capacity for a solar power system involves considering various factors related to electricity consumption, solar potential, and energy goals. This is a step-by-step approach to help you estimate the appropriate system size and capacity:

1. Determine your electricity consumption: Start by analyzing your historical electricity bills and load profiling data to understand your average and peak electricity consumption. Identify your annual kilowatt-hour (kWh) usage, as this will be a key factor in sizing the solar system.

2. Assess your energy goals: Determine your energy goals, such as offsetting a certain percentage of your electricity usage, achieving energy independence, or maximizing financial savings. This will guide the system size and capacity you aim to achieve.

3. Evaluate your solar potential: Assess the solar potential of your site by considering factors like roof orientation, tilt, shading, available space, and local solar irradiance. You can consult solar professionals or use online tools to estimate the solar energy generation potential based on your specific location.

4. Calculate system size based on energy consumption: To estimate the system size, divide your annual kWh consumption by the average annual solar energy production per installed kilowatt (kW) of solar capacity. The average solar energy production per kW varies depending on factors like location, climate, and system efficiency. A rule of thumb is that a well-designed solar system can produce around 1,200 to 1,500 kWh per installed kW per year. Adjust this value based on local conditions or consult with solar professionals for a more accurate estimation. System Size (kW) = Annual kWh Consumption / Average Annual Solar Energy Production per kW

5. Consider system capacity and panel efficiency: Take into account the efficiency of the solar panels you plan to install. Higher-efficiency panels can generate more electricity in a given area, allowing you to achieve a desired system capacity with fewer panels. The system capacity is the total maximum output capacity of the solar system and is typically expressed in kilowatts (kW) or megawatts (MW).

6. Account for system losses: It's important to consider system losses due to factors like shading, panel soiling, wiring losses, and inverter efficiency. Deducting a certain percentage (usually around 10-20%) from the estimated system size can help account for these losses and ensure the desired energy production is achieved.

7. Assess budget and available space: Consider your budget and available space for installing solar panels. Balance the desired system size with practical constraints to ensure a feasible and cost-effective installation.

8. Consult with solar professionals: Seeking advice from solar professionals or engineers can provide valuable insights and expertise in estimating the system size and capacity. They can conduct a detailed site assessment, shading analysis, and energy modeling to help you make informed decisions about system sizing.

CONSIDERING FUTURE ENERGY NEEDS

Considering future energy needs is an important aspect when planning a solar power system. Anticipating changes in energy consumption patterns and incorporating future energy needs in the system design can help ensure the long-term viability and adequacy of the solar system. Some factors to consider when evaluating future energy needs:

1. Lifestyle Changes: Assess any planned or anticipated lifestyle changes that may affect your energy consumption. For example, if you are planning to expand your family, purchase electric vehicles, install new appliances, or add energy-intensive equipment or systems, these changes will increase your electricity demands. Consider the energy requirements of these future additions and factor them into your system size estimation.

2. Energy Efficiency Improvements: Implementing energy efficiency measures before installing a solar system can help reduce your overall energy consumption. Consider potential energy-saving upgrades such as improving insulation, upgrading to energy-efficient appliances and lighting, installing smart home technologies, or implementing energy management systems. By reducing your baseline energy consumption, you can optimize the system size and capacity required to meet your future energy needs.

3. Net Metering and Time-of-Use Rates:
Understand the net metering policies and time-of-use (TOU) rates offered by your utility company. Net metering allows you to feed excess solar energy back to the grid and receive credits for it, which can be used during periods of lower solar production or higher energy demand. TOU rates vary the cost of electricity depending on the time of day, incentivizing the consumption of solar energy during peak-rate periods. By aligning your system design and energy usage strategy with

these policies, you can optimize the financial benefits and balance your energy needs.

4. Energy Storage: Consider incorporating energy storage solutions, such as batteries, into your solar power system design. Energy storage allows you to store excess solar energy generated during the day and use it during times when solar production is low or during power outages. Adding energy storage provides flexibility, enhances self-consumption, and allows for greater energy independence. Assess your future energy needs and determine whether energy storage is a viable option to meet those needs.

5. System Scalability: Design your solar system with scalability in mind. Consider the potential for future expansion or the addition of more solar panels as your energy needs grow. Consult with solar professionals who can provide guidance on system design that allows for easy integration of additional panels or capacity expansion.

6. Financial Planning: Evaluate the financial implications of future energy needs and system expansion. Assess the return on investment (ROI) and payback period for the initial solar installation, as well as any potential future investments in system expansion or energy storage. Consider the availability of financing options, incentives, and tax credits that can help support your future energy plans.

CHAPTER IV. CHOOSING THE RIGHT SOLAR SYSTEM

TYPES OF SOLAR INSTALLATIONS

There are several types of solar installations, each with its own characteristics and applications. These are some common types:

1. Rooftop Solar Installations: Rooftop solar installations are the most common type of solar system. They involve mounting solar panels on the rooftops of residential, commercial, or industrial buildings. Rooftop installations take advantage of available space and are suitable for properties with sufficient sunlight exposure. They can be either grid-tied, where excess energy is fed back to the grid, or off-grid, where energy is stored in batteries for self-consumption.

2. Ground-Mounted Solar Installations: Ground-mounted solar installations involve mounting solar panels on open land or on dedicated solar farms. This type of installation is ideal for large-scale solar projects, where there is ample available space and minimal shading. Ground-mounted systems can be easily optimized for the best sun exposure and can accommodate a higher number of solar panels compared to rooftop installations.

3. Solar Carport Installations: Solar carports combine the benefits of solar energy generation with the provision of shade and weather protection for parked vehicles. Solar panels are installed on the top of the carport structure, allowing for dual-use of space. Solar carports are commonly found in parking lots, shopping centers, and office complexes, providing renewable energy and reducing the carbon footprint of transportation.

4. Building-Integrated Photovoltaics (BIPV): BIPV installations involve integrating solar panels directly into the building materials, such as solar roof tiles, solar windows, or solar facades. BIPV systems serve both as functional building elements and power generators. They offer aesthetic appeal and can be seamlessly incorporated into new construction or retrofitted into existing structures.

5. Floating Solar Installations: Floating solar installations are mounted on bodies of water, such as lakes, reservoirs, or ponds. Floating solar arrays utilize water surface area to generate electricity. They can be advantageous in areas where land availability is limited or in locations where water bodies are abundant. Floating solar installations can also help reduce water evaporation and provide cooling benefits to the solar panels.

6. Solar Tracking Systems: Solar tracking systems automatically orient solar panels to track the movement of the sun throughout the day. These systems maximize solar energy generation by continuously adjusting the panel angles to optimize sun exposure. Solar tracking systems can be used in various installations, including rooftop, ground-mounted, and solar farms.

7. Off-Grid Solar Installations: Off-grid solar installations are designed to operate independently of the electrical grid. They are commonly used in remote areas where grid connection is not available or in situations where energy independence is desired. Off-grid systems typically incorporate batteries or other energy storage solutions to store excess energy for use during periods of low solar production.

GRID-TIED SYSTEMS

Grid-tied solar systems, also known as grid-connected systems, are solar installations that are connected to the electrical grid. They work in conjunction with the utility grid, allowing the exchange of electricity between the solar system and the grid. Some key features and benefits of grid-tied solar systems:

1. Power Generation and Grid Interaction: Grid-tied systems generate electricity from solar panels and supply it to the building or premises where they are installed. If the solar system produces more electricity than is currently needed, the excess power can be fed back into the grid. When the solar system is not generating enough electricity to meet the demand, electricity can be drawn from the grid.

2. Net Metering: One of the significant advantages of grid-tied systems is the availability of net metering programs in many regions. Net metering allows homeowners or businesses to receive credits for excess solar energy they generate and feed back into the grid. These credits can then be used to offset the electricity drawn from the grid during times when the solar system is not producing enough energy, such as at night or during cloudy periods. Net metering effectively enables the homeowner or business to "bank" their excess energy and use it when needed.

3. Financial Benefits: Grid-tied systems offer financial benefits through net metering programs. By offsetting or reducing electricity bills, grid-tied solar systems can lead to significant cost savings over time. In some cases, if the solar system produces more electricity than is consumed over a billing cycle, homeowners or businesses may receive a check or credit from the utility company for the excess energy generated.

4. Grid Stability and Reliability: Grid-tied systems contribute to the stability and reliability of the electrical grid. During peak solar production periods, when solar systems generate excess electricity, the grid benefits from the additional power supply. Conversely, when solar production is low, such as at night, grid electricity is readily available for consumption. This balanced exchange helps maintain a stable grid and reduces the strain on the electrical infrastructure.

5. Easy Installation and Flexibility: Grid-tied systems are relatively straightforward to install compared to off-grid systems since they don't require the use of batteries or complex energy storage systems. The absence of batteries also means there is less maintenance and fewer components to manage. Grid-tied systems can be installed on rooftops, ground-mounted arrays, or incorporated into various structures, offering flexibility in design and placement.

6. Environmental Benefits: By generating electricity from solar energy, grid-tied systems reduce reliance on fossil fuel-based electricity generation. This helps reduce greenhouse gas emissions and contributes to a cleaner and more sustainable energy future. Grid-tied solar systems make a positive environmental impact by promoting the use of renewable energy sources.

OFF-GRID SYSTEMS

Off-grid solar systems, also known as standalone solar systems, are independent energy systems that operate without a connection to the electrical grid. They are typically used in remote areas where grid connection is not available or in situations where energy independence is desired. Here are some key features and benefits of off-grid solar systems:

1. Energy Independence: Off-grid systems allow homeowners, businesses, or communities to be self-sufficient in terms of electricity generation. They provide a reliable and independent power source, ensuring uninterrupted electricity supply even in areas without access to the grid. This is particularly beneficial in remote locations or areas prone to power outages.

2. Battery Storage: Off-grid systems incorporate batteries or other energy storage solutions to store excess energy generated during the day for use during times of low solar production, such as at night or during cloudy weather. The stored energy in the batteries ensures a continuous power supply even when solar energy generation is limited.

3. Standalone Power Generation: Off-grid systems rely solely on solar energy to generate electricity. They typically consist of solar panels, charge controllers, batteries, inverters, and sometimes backup generators. These components work together to convert solar energy into usable electricity, store it in batteries, and convert it back to AC power for household or commercial use.

4. Energy Management and Efficiency: Off-grid systems require careful energy management and efficiency practices to ensure the available energy is used wisely. Users need to be mindful of their energy consumption and make conscious decisions to optimize energy usage. Energy-efficient appliances, LED lighting, and smart energy management systems can further enhance the efficiency of off-grid systems.

5. Remote Power Applications: Off-grid systems are commonly used in remote locations where connecting to the grid is impractical or cost-prohibitive. They are suitable for applications such as cabins, remote homes, agricultural operations, telecommunications towers, and scientific research stations in isolated areas. Off-grid systems can provide essential electricity for lighting, appliances, water pumping, refrigeration, and other basic electrical needs.

6. Environmental Benefits: Off-grid solar systems rely on renewable energy sources, such as sunlight, to generate electricity. By reducing or eliminating the reliance on fossil fuel-based generators, off-grid systems help minimize greenhouse gas emissions and contribute to a cleaner and more sustainable energy future. They provide an environmentally friendly alternative to traditional diesel or gasoline generators used in off-grid areas.

7. Customization and Scalability: Off-grid systems can be tailored to meet specific energy needs and requirements. They can be designed and sized based on the desired energy consumption, available solar resources, and the storage capacity needed. Additionally, off-grid systems can be expanded or modified over time as energy needs change or as additional power sources, such as wind turbines or micro-hydro systems, are integrated.

HYBRID SYSTEMS

Hybrid solar systems, also known as hybrid renewable energy systems, combine multiple sources of energy generation to provide a reliable and efficient power supply. These systems typically integrate solar energy with other renewable energy sources, such as wind turbines or micro-hydro systems, along with energy storage solutions. Some key features and benefits of hybrid systems:

1. Power Generation Optimization: Hybrid systems leverage the strengths of different renewable energy sources to maximize power generation. Solar panels are highly effective during the day, while wind turbines or micro-hydro systems can generate electricity even during periods of low solar production. By combining these sources, hybrid systems ensure a more consistent and reliable power supply throughout the day and under varying weather conditions.

2. Energy Storage and Management: Hybrid systems incorporate energy storage solutions, such as batteries, to store excess energy generated by solar panels or other renewable sources. The stored energy can be used during periods of low energy production or during the night. Effective energy management and storage ensure a steady power supply and help balance the energy needs with the available renewable energy resources.

3. Grid Interaction and Backup Power: Hybrid systems can be connected to the electrical grid, allowing for grid interaction and the ability to sell excess energy back to the grid through net metering programs. In the event of a grid outage, hybrid systems with energy storage can automatically switch to backup power mode, providing continuous electricity to critical loads or the entire premises. This feature is particularly beneficial in areas with unreliable grid infrastructure or where backup power is essential.

4. Flexibility and Scalability: Hybrid systems offer flexibility and scalability in terms of design and capacity. They can be customized based on specific energy needs, available renewable resources, and the desired level of energy independence. Hybrid systems can be expanded over time by adding additional solar panels, wind turbines, or energy storage capacity to meet growing energy demands.

5. Cost Savings and Environmental Benefits: Hybrid systems help reduce reliance on grid electricity and fossil fuel-based power generation. By incorporating renewable energy sources, they contribute to lower electricity bills and reduce greenhouse gas emissions. Hybrid systems offer both financial savings through reduced energy costs and environmental benefits by promoting the use of clean and sustainable energy sources.

6. Remote Power Applications: Hybrid systems are particularly useful in remote areas where grid connection is unavailable or unreliable. They provide a reliable and independent power supply for remote homes, off-grid cabins, telecommunications towers, or agricultural operations. Hybrid systems can help meet the energy needs of these remote applications while minimizing reliance on diesel generators and their associated costs and environmental impacts.

SELECTING SOLAR PANELS

Different types and technologies

There are various types of solar technologies and systems available today. Some of the common ones:

1. Monocrystalline Silicon (Mono-Si): Monocrystalline solar panels are made from a single crystal structure, typically of high-purity silicon. They have a uniform black appearance and high efficiency, making them one of the most popular and efficient solar panel technologies available. Monocrystalline panels are known for their high power output and good performance in limited space.

2. Polycrystalline Silicon (Poly-Si):

Polycrystalline solar panels are made from multiple silicon crystals. They have a textured appearance with a blue hue. Polycrystalline panels are cost-effective and offer good efficiency. While they may have slightly lower efficiency compared to monocrystalline panels, they are widely used due to their affordability.

3. Thin-Film Solar Cells: Thin-film solar cells are made by depositing layers of photovoltaic material onto a substrate, such as glass or metal. Thin-film technology includes various types such as amorphous silicon (a-Si), cadmium telluride (CdTe), and copper indium gallium selenide (CIGS). Thin-film solar cells are lightweight, flexible, and can be produced in large quantities. They are suitable for certain applications where weight, flexibility, or aesthetic considerations are important.

4. Concentrated Solar Power (CSP):

Concentrated solar power systems use mirrors or lenses to concentrate sunlight onto a receiver, which converts the heat into electricity. CSP technology is often used in large-scale power plants and can include different configurations such as parabolic troughs, power towers, and dish/engine systems. CSP systems are capable of storing heat for continuous power generation even when the sun is not shining, offering a form of thermal energy storage.

5. Solar Thermal Collectors: Solar thermal collectors capture sunlight and convert it into heat energy for applications such as water heating, space heating, or industrial processes. Solar thermal systems use different collector types, including flat-plate collectors, evacuated tube collectors, and parabolic troughs. The collected heat can be used directly or stored in insulated containers or heat storage systems for later use.

6. Building-Integrated Photovoltaics (BIPV):
BIPV systems integrate solar panels directly into building materials, such as solar roof tiles, solar windows, or solar facades. BIPV technology serves both as a functional building element and a power generator. It offers aesthetic appeal, seamless integration into the building's design, and potential cost savings by reducing the need for conventional building materials.

7. Solar Tracking Systems: Solar tracking systems automatically orient solar panels to track the movement of the sun throughout the day, maximizing solar energy generation. There are two main types of solar tracking systems: single-axis and dual-axis trackers. Single-axis trackers move the panels on one axis (usually east to west), while dual-axis trackers adjust the panels on both horizontal and vertical axes, optimizing the panel's angle to the sun's position.

EFFICIENCY AND PERFORMANCE CONSIDERATIONS

When considering solar energy systems, efficiency and performance are crucial factors to evaluate. Some key considerations related to efficiency and performance:

1. Solar Panel Efficiency: The efficiency of solar panels refers to their ability to convert sunlight into electricity. Higher efficiency panels can generate more power for a given amount of sunlight. Monocrystalline panels generally have higher efficiency compared to polycrystalline and thin-film panels. When evaluating solar panels, it's important to consider both the initial efficiency and the degradation rate over time.

2. System Performance Ratio (SPR): The System Performance Ratio is a metric that evaluates the overall performance of a solar energy system. It takes into account various factors, including the efficiency of the solar panels, losses in the system due to wiring, shading, and other factors. A higher SPR indicates a more efficient system.

3. Temperature Coefficient: Solar panels can experience a decrease in efficiency as their temperature rises. The temperature coefficient indicates the rate of efficiency decline with increasing panel temperature. Panels with lower temperature coefficients perform better in hot climates.

4. Shading Considerations: Shading can significantly impact the performance of solar panels. Even partial shading on a small portion of a solar panel can cause a disproportionate decrease in overall system output. It's important to assess shading patterns throughout the day and year and take steps to minimize shading, such as panel placement or trimming trees.

5. Orientation and Tilt: The orientation and tilt angle of solar panels affect their exposure to sunlight. In most regions, optimal orientation is facing south (in the Northern Hemisphere) or north (in the Southern Hemisphere) to maximize solar energy capture. The tilt angle is typically set to match the latitude of the installation site. Proper orientation and tilt can significantly enhance system performance.

6. Energy Monitoring and Maintenance: Monitoring the energy production of your solar system allows you to track its performance over time. Energy monitoring systems provide real-time data on energy generation, allowing you to identify any issues or inefficiencies. Regular maintenance, such as cleaning panels and inspecting connections, is also important to ensure optimal performance.

7. Inverter Efficiency: Inverters convert the DC electricity generated by solar panels into AC electricity for use in homes or businesses. The efficiency of the inverter affects the overall system efficiency. High-quality inverters with higher efficiency ratings help minimize energy losses during the conversion process.

8. System Sizing and Design: Proper system sizing is essential to achieve optimal performance. An accurately sized system takes into account the energy consumption needs, available sunlight, and other factors to ensure the system generates enough power to meet the demand. Oversized or undersized systems can lead to inefficiencies and suboptimal performance.

EVALUATING WARRANTIES AND CERTIFICATIONS

When considering solar energy systems, it's important to evaluate warranties and certifications associated with the components and the installation. Some key aspects to consider:

1. Panel Warranty: Solar panels typically come with a warranty that guarantees their performance and durability. The warranty usually consists of two components: a performance warranty and a product warranty. The performance warranty guarantees a certain level of power output over a specified period (e.g., 25 years), while the product warranty covers defects and issues with the panel itself. Look for panel warranties with longer durations and favorable terms.

2. Inverter Warranty: Inverters also come with warranties that cover their performance and durability. The warranty period for inverters can vary, but a standard warranty period is around 5 to 10 years. Some manufacturers offer extended warranty options for an additional cost. It's important to understand the warranty coverage and any limitations or exclusions.

3. Installation Warranty: The solar installation itself may come with a warranty provided by the installer or the solar company. This warranty typically covers the quality of the installation workmanship and ensures that the system is installed correctly. It's important to review the terms and duration of the installation warranty and understand the responsibilities of the installer in case of any issues.

4. Certifications and Standards: Look for solar panels and components that have been certified by reputable organizations and meet recognized industry standards. Some certifications to look for include:

- **International Electrotechnical Commission (IEC) certification:** This ensures compliance with international standards for safety and performance.

- **Underwriters Laboratories (UL) certification:** UL certification ensures that the panels meet specific safety and performance standards.

- **TÜV certification:** TÜV certification is a widely recognized certification for product safety and reliability.

5. Company Reputation: Consider the reputation and track record of the solar panel manufacturer, inverter manufacturer, and the installation company. Look for companies with a proven history of delivering quality products and services.

Online reviews, customer testimonials, and industry recognition can provide insights into the reputation and reliability of the companies involved.

6. Extended Warranty Options: Some manufacturers or installers may offer extended warranty options for an additional cost. These extended warranties can provide additional coverage beyond the standard warranty period and offer peace of mind.

INVERTERS AND BALANCE OF SYSTEM COMPONENTS

Inverter types (string vs. micro inverters)

When it comes to solar energy systems, there are two primary types of inverters: string inverters and micro inverters.

1. String Inverters: String inverters, also known as central inverters, are the traditional and most commonly used type of inverters in solar installations. They are installed at a centralized location and are connected to multiple solar panels wired together in series, forming a string. The DC power generated by the solar panels is sent to the string inverter, which converts it into AC power for use in homes or businesses.

Key Features and Considerations:

Cost-Effective: String inverters are generally more cost-effective compared to micro inverters.

Single Point of Failure: Since string inverters are centralized, the entire system's performance can be affected if one panel or a section of panels experiences shading, dirt, or other issues.

Design Flexibility: String inverters are well-suited for larger installations where there is ample roof space and consistent panel orientation.

Monitoring and Maintenance: String inverters typically provide monitoring capabilities to track the overall system's performance. Maintenance and troubleshooting are centralized, as the inverter is easily accessible for inspections or repairs.

2. Micro inverters: Micro inverters, on the other hand, are installed on each individual solar panel in the system. Instead of wiring the panels in series, each panel is equipped with its own micro inverter. Each micro inverter converts the DC power generated by its respective panel into AC power.

Key Features and Considerations:

Panel-Level Optimization: Micro inverters enable panel-level optimization, which means that each panel operates independently, maximizing the energy production of each panel, regardless of shading or soiling on other panels. This leads to higher overall system efficiency and performance.

Increased Flexibility: Micro inverters offer flexibility in system design and installation. Panels can be installed on different roof orientations or even on multiple roofs without compromising the system's performance.

Monitoring and Maintenance: Micro inverters often come with built-in monitoring capabilities, allowing for real-time monitoring of each panel's performance. If a panel or micro inverter experiences an issue, it can be easily identified and addressed without impacting the performance of other panels.

Higher Costs: Micro inverters are generally more expensive compared to string inverters due to the additional components required for each panel.

Longevity: Micro inverters are typically designed to last as long as the solar panel itself. In case of a failure, only the affected panel is impacted, minimizing any downtime or loss of system performance.

The choice between string inverters and micro inverters depends on various factors, including the size of the installation, shading considerations, design flexibility, monitoring preferences, and budget. It's recommended to consult with solar professionals or installers who can assess your specific needs and provide guidance on the most suitable inverter type for your solar energy system.

CHARGE CONTROLLERS AND BATTERIES (FOR OFF-GRID SYSTEMS)

In off-grid solar systems, charge controllers and batteries play a crucial role in managing the storage and utilization of electrical energy. Here's an overview of charge controllers and batteries in off-grid systems:

1. Charge Controllers:

Charge controllers, also known as charge regulators, are devices that regulate the flow of electrical current between the solar panels and the batteries. Their primary function is to prevent overcharging and over-discharging of the batteries, ensuring their longevity and optimal performance.

Key Features and Considerations:

Charging Regulation: Charge controllers monitor the battery voltage and adjust the charging current to prevent overcharging. They ensure that the batteries receive the appropriate charge and protect them from damage due to excessive voltage.

Battery Protection: Charge controllers also prevent over-discharging of the batteries, which can lead to reduced battery capacity and lifespan. They disconnect the load from the batteries when the voltage drops to a certain threshold, preserving the battery's energy. Types of Charge Controllers: There are primarily two types of charge controllers: PWM (Pulse Width Modulation) and MPPT (Maximum Power Point Tracking). PWM controllers are simpler and more affordable but have lower efficiency compared to MPPT controllers. MPPT controllers are more sophisticated and can extract maximum power from the solar panels, resulting in higher energy harvest.

2. Batteries:

Batteries are essential components of off-grid solar systems as they store excess energy generated by the solar panels for use during periods of low or no sunlight. They provide a reliable source of electricity when the solar panels are not actively generating power.

Key Features and Considerations:

Battery Types: Common battery types used in off-grid systems include lead-acid batteries (flooded, gel, and AGM) and lithium-ion batteries. Lead-acid batteries are cost-effective but require regular maintenance, while lithium-ion batteries are more expensive but offer higher energy density and longer lifespan.

Capacity and Voltage: The battery capacity is measured in ampere-hours (Ah) and determines the amount of energy it can store. The battery voltage, usually 12V, 24V, or 48V, should be compatible with the system's voltage requirements.

Depth of Discharge (DoD): The depth of discharge refers to the percentage of a battery's total capacity that is discharged before recharging. Lead-acid batteries typically have a higher recommended DoD (around 50%) compared to lithium-ion batteries (usually up to 80%). Deeper discharge can affect battery lifespan, so it's important to adhere to the recommended DoD.

Battery Maintenance: Lead-acid batteries require regular maintenance, including checking electrolyte levels, equalizing charges, and ensuring proper ventilation. Lithium-ion batteries are generally maintenance-free.

Lifespan and Warranty: Batteries have a limited lifespan, typically stated in terms of cycles (number of charge and discharge cycles) or years of service. Consider the expected lifespan and warranty of the battery when making a selection.

SAFETY FEATURES AND PROTECTION DEVICES

Safety features and protection devices are essential components of solar energy systems to ensure the safe and reliable operation of the system. Some important safety features and protection devices commonly found in solar installations:

1. Disconnect Switches: Disconnect switches are devices that allow the system to be safely disconnected from the electrical grid or other power sources. They provide a means to isolate the solar panels, inverters, and batteries from the rest of the electrical system during maintenance, repairs, or emergencies. Disconnect switches are typically installed on both the AC and DC sides of the system.

2. Surge Protection Devices (SPDs): SPDs, also known as surge suppressors or surge arrestors, protect the system from voltage surges caused by lightning strikes, utility grid fluctuations, or other transient events. They divert excessive voltage spikes to the ground, preventing damage to sensitive components like inverters, charge controllers, and electronic devices.

3. Ground Fault Protection: Ground fault protection devices monitor the electrical currents flowing through the system and detect any imbalances or leakage of current to ground. They quickly disconnect the system in case of a ground fault, reducing the risk of electric shock and fire hazards. Ground fault protection is particularly important in systems with exposed metal surfaces or in locations where there is a higher risk of moisture or water contact.

4. Overcurrent Protection: Overcurrent protection devices, such as fuses and circuit breakers, are used to protect the electrical wiring, components, and equipment from excessive current flow. They are installed in series with the wiring and are designed to trip or open the circuit when the current exceeds the rated capacity. Overcurrent protection devices prevent overheating and potential fire hazards due to overloading or short circuits.

5. Fire Safety Measures: Fire safety is a critical consideration in solar installations. Various measures can be implemented to mitigate fire risks, including fire-rated wiring, appropriate spacing between panels, fire-resistant materials for roof penetrations, and proper ventilation for battery storage areas. Fire extinguishers and smoke detectors should also be installed as part of the overall fire safety plan.

6. Electrical Codes and Compliance: Solar installations must comply with relevant electrical codes and regulations to ensure safety. Compliance with codes such as the National Electrical Code (NEC) or local building codes helps ensure proper installation practices, equipment selection, and system grounding.

7. Warning Labels and Signage: Clear and visible warning labels and signage should be placed at key locations within the solar system to alert individuals to potential hazards, electrical risks, and safety precautions. These labels provide important information regarding high voltage, arc flash hazards, and proper procedures for working on the system.

CHAPTER V. INSTALLING YOUR SOLAR POWER SYSTEM

HIRING PROFESSIONALS VS. DIY INSTALLATION

Deciding between hiring professionals or undertaking a DIY installation for a solar energy system depends on several factors. Here are some factors to consider that can assist you in making an informed decision:

1. Technical Expertise: Solar energy systems involve electrical wiring, equipment installation, and system integration. Professional installers have the technical expertise and knowledge to design, install, and commission a system correctly. They are familiar with local building codes, safety requirements, and best practices. If you lack experience or knowledge in electrical work and solar installations, hiring professionals is recommended to ensure a safe and efficient system.

2. Safety: Safety is of utmost importance when

working with electrical components and high-voltage systems. Professional installers are trained in safety protocols and have the necessary tools and equipment to handle installations safely. They are aware of potential risks, such as electrical shocks, fire hazards, and fall hazards, and know how to mitigate them. DIY installations may pose safety risks if proper precautions are not taken.

3. System Design and Sizing: Designing a solar energy system requires evaluating various factors such as site suitability, energy consumption, panel orientation, shading analysis, and equipment selection. Professional installers have the expertise to assess these factors accurately and design a system that meets your specific needs. They can help optimize the system's performance and ensure it is properly sized for your energy requirements.

4. Permits and Paperwork: Solar installations often require permits and approvals from local authorities. Professional installers are familiar with the permit process and can handle the paperwork on your behalf. They understand the requirements and can ensure compliance with building codes and regulations. DIY installations may require you to navigate the permitting process yourself, which can be time-consuming and challenging.

5. Warranty and Support: Professional installers often provide warranties on their workmanship and the components used. This means you have recourse in case of any issues or defects in the system. They also offer ongoing support and maintenance services, ensuring the system operates optimally. DIY installations may not come with the same level of warranty coverage and ongoing support.

6. Time and Effort: Installing a solar energy system can be complex and time-consuming, especially for someone without experience. It involves tasks such as rooftop work, electrical wiring, mounting, and system testing. DIY installations require a significant investment of time, effort, and research to ensure a successful outcome. Consider whether you have the availability and commitment to undertake such a project.

7. Cost: While DIY installations may appear cost-effective at first glance, there are several factors to consider. Professional installers often have access to discounted equipment pricing, industry connections, and knowledge of available incentives and rebates. They can help you maximize financial benefits and avoid costly mistakes. Additionally, DIY installations may not be eligible for certain warranties, incentives, or insurance coverage.

FINDING QUALIFIED SOLAR INSTALLERS

Finding qualified solar installers for your solar energy project is essential to ensure a professional and reliable installation. Some ways to find qualified solar installers:

1. Recommendations: Seek recommendations from friends, family, or acquaintances who have previously installed solar energy systems. Their firsthand experiences can provide valuable insights and help you identify reputable installers.

2. Online Directories: Utilize online directories that list qualified solar installers in your area. Websites like the Solar Energy Industries Association (SEIA), North American Board of Certified Energy Practitioners (NABCEP), and local solar energy associations often have directories or search tools to find certified installers.

3. Online Reviews and Ratings: Read online reviews and ratings for solar installers in your area. Websites like Yelp, Google Reviews, and Angie's List can provide insights into the experiences of previous customers. Consider both positive and negative feedback to make an informed decision.

4. Local Solar Energy Associations: Check with local solar energy associations or renewable energy organizations in your area. They may have resources, lists, or directories of certified installers who are members of the association.

5. Certification and Credentials: Look for solar installers who hold certifications from reputable organizations like NABCEP. NABCEP certification signifies that the installer has met rigorous standards of knowledge and experience in solar system installation.

6. Multiple Quotes and Evaluations: Obtain quotes and estimates from multiple installers. This will give you a better understanding of the market rates, equipment options, and installation approaches. Compare the quotes, evaluate the proposed system designs, and consider the reputation and experience of the installers before making a decision.

OBTAINING NECESSARY PERMITS AND APPROVALS

Obtaining the necessary permits and approvals for your solar energy system is a crucial step to ensure compliance with local regulations and building codes. The specific permitting process can vary depending on your location, so it's important to consult with your local authorities and follow their guidelines.

1. Research Local Requirements: Begin by researching the specific requirements for solar installations in your area. Contact your local building department, planning/zoning department, or permit office to inquire about the necessary permits, codes, and regulations that apply to solar energy systems. They can provide you with the specific guidelines and documentation needed for your project.

2. Gather Documentation: Prepare the documentation required for the permit application. This typically includes the system design plans, equipment specifications, electrical diagrams, structural engineering reports (if applicable), and any other supporting documentation requested by the authorities. Your solar installer or system designer can assist you in preparing these documents.

3. Complete Permit Application: Fill out the permit application form provided by the local authorities. Include all the necessary information, such as property details, system specifications, contractor information, and contact details. Ensure accuracy and completeness to avoid delays in the review process.

4. Application Review: Submit the completed permit application along with the required documentation to the relevant department or office. The authorities will review your application to ensure compliance with local regulations, building codes, and safety standards. They may request revisions or additional information during the review process.

5. Fees and Payments: Pay any applicable permit fees associated with the application. The costs associated with solar power systems can fluctuate based on factors such as the system's size and intricacy. Check with the local authorities for the accepted payment methods and fee structure.

6. Inspection: Once the permit is approved, schedule inspections at different stages of the installation process as required by the authorities. Common inspection points include pre-installation, wiring, and final inspections. Ensure that the system is installed according to the approved plans and complies with safety and electrical codes.

7. Final Approval: After successful inspections, the authorities will issue a final approval or certificate of completion, indicating that your solar energy system has been installed and inspected according to the requirements. Keep this documentation for your records.

ASSESSING STRUCTURAL INTEGRITY AND ELECTRICAL CAPACITY

Assessing the structural integrity and electrical capacity of your property is an important step when considering a solar energy system installation.

Structural Integrity:

1. Roof Evaluation: Determine if your roof is structurally sound and capable of supporting the weight of solar panels. Consider factors such as the age, condition, and type of roofing material. It's advisable to have a professional roofing inspection to assess its integrity and identify any repairs or reinforcements needed.

2. Load-Bearing Capacity: Consult with a structural engineer or a professional solar installer to evaluate the load-bearing capacity of your roof. They will consider factors such as the roof structure, material strength, and local snow and wind loads. This assessment ensures that your roof can safely accommodate the additional weight of the solar panels.

3. Mounting Options: Assess the mounting options available for your property, such as roof-mounted, ground-mounted, or pole-mounted systems. The suitability of each option depends on factors like available space, sun exposure, local regulations, and aesthetic preferences.

4. Ground Conditions: If you are considering ground-mounted systems, evaluate the ground conditions to ensure stability and proper anchoring. Factors such as soil type, slope, and drainage should be considered. In some cases, soil testing and site preparation may be necessary.

ELECTRICAL CAPACITY:

1. Electrical Panel Evaluation: Assess the capacity and condition of your existing electrical panel (also known as the main service panel or distribution board). Determine if it has sufficient space and capacity to accommodate the additional solar energy system. An electrician can help evaluate the panel and determine if an upgrade or replacement is necessary.

2. Electrical Load Analysis: Evaluate your household's electricity consumption patterns to estimate the system size needed. Consider the average monthly and annual energy usage, peak load, and any planned changes in energy consumption. This analysis will help determine the appropriate capacity and design of the solar energy system.

3. Electrical Upgrades: If your electrical panel or wiring is outdated or inadequate, you may need to upgrade or expand your electrical system. This can involve installing a larger electrical panel, upgrading wiring, or adding additional circuits to support the solar system's electrical requirements. An electrician should be consulted to ensure compliance with electrical codes and safety standards.

MOUNTING SOLAR PANELS

Mounting solar panels properly is essential to ensure their stability, optimize energy production, and maintain the integrity of your roof or ground structure. The basic steps involved in mounting solar panels:

1. System Design: Before mounting the panels, ensure you have a well-designed system layout that considers factors like the available space, sun exposure, shading analysis, and local regulations. The design should also account for the specific mounting method chosen.

2. Roof Mounting:

- **Roof Inspection:** Conduct a thorough inspection of your roof to identify any existing damage, leaks, or structural issues. It's recommended to address any necessary repairs or maintenance before proceeding with the solar panel installation.

- **Racking System:** Install a racking system designed for solar panel mounting. Racking systems typically include rails, brackets, and clamps that provide a secure and adjustable framework for the panels. Follow the manufacturer's instructions and guidelines for proper installation.

- **Flashing and waterproofing:** Install flashing and waterproofing components around the roof penetrations, such as roof attachments and attachment points for the rails. This helps prevent water leaks and ensures the long-term integrity of your roof.

- **Panel Installation:** Attach the solar panels to the racking system using the specified mounting hardware. Follow the manufacturer's instructions and recommendations for panel alignment, spacing, and tightening torque.

- **Electrical Connections:** Once the panels are securely mounted, make the necessary electrical connections, such as wiring the panels in series or parallel, and connecting them to the inverter or junction box.

3. Ground Mounting:

- **Site Preparation:** If you're installing ground-mounted solar panels, prepare the installation site by clearing any obstructions, leveling the ground, and addressing any soil concerns.

- **Foundation or Ground Screws:** Depending on the chosen mounting method, install the appropriate foundation or ground screws to anchor the mounting structure securely into the ground. Follow the manufacturer's instructions for proper installation techniques.

- **Racking System:** Similar to roof-mounted systems, install a racking system suitable for ground-mounted solar panels. This will provide the necessary support and adjustability for panel placement.

- **Panel Installation:** Attach the solar panels to the racking system, ensuring they are properly aligned, spaced, and tightened according to the manufacturer's specifications.

- **Electrical Connections:** Connect the panels to the necessary electrical components, such as the wiring system, combiner box, or inverter, following electrical codes and safety guidelines.

4. Safety Precautions: During the installation process, it's crucial to follow safety protocols. Use appropriate safety equipment, such as harnesses, ladders, and fall protection, when working at heights. If you're not experienced with roof work or electrical connections, consider hiring professionals to ensure the installation is done safely and efficiently.

It's important to consult the manufacturer's instructions, local building codes, and regulations specific to your area throughout the installation process. If you're unsure about any aspect of mounting the solar panels, it's recommended to seek guidance from a professional solar installer or a qualified contractor who has experience in solar panel installations.

CONNECTING WIRING AND COMPONENTS

Connecting wiring and components in a solar energy system is a critical step to ensure the proper functioning and safety of the system. The key steps involved in connecting the wiring and components:

1. Solar Panels:

- **Connect the solar panels together:** If you have multiple solar panels, connect them in series or parallel configuration based on the system design. Series connections result in an increase in voltage, while parallel connections lead to an increase in current.

- Use appropriately sized and rated DC-rated cables to connect the positive (+) and negative (-) terminals of the solar panels.

- **Ensure proper polarity:** Connect the positive terminal of one panel to the negative terminal of the next panel, and continue until all panels are connected.

2. Combiner Box:

The combiner box is used to consolidate the output of multiple solar panel strings into a single set of wires for easier routing to the inverter or charge controller.

- Connect the output cables from each solar panel string to the appropriate terminals in the combiner box.

- Ensure proper fusing: Install fuses or circuit breakers in the combiner box for overcurrent protection as per the system requirements and local electrical codes.

3. Inverter or Charge Controller:

- Connect the output of the combiner box or solar panels directly to the input terminals of the inverter or charge controller, depending on the type of system (grid-tied or off-grid).
- Follow the manufacturer's instructions for proper wiring connections, ensuring proper polarity and tightening of connections.
- Use appropriately sized and rated DC cables for the connections between the solar panels and the inverter or charge controller.

4. Electrical Panel:

- If you have a grid-tied system, connect the output of the inverter to the main electrical panel of your property.
- Install a dedicated circuit breaker or disconnect switch for the solar energy system in the electrical panel. This allows you to shut off power to the system when necessary.

- Follow local electrical codes and regulations when making connections to the electrical panel. It's recommended to involve a licensed electrician to ensure compliance and safety.

5. Monitoring and Control Devices:

- If your system includes monitoring and control devices, such as energy meters, data loggers, or smart home integration, follow the manufacturer's instructions for their installation and wiring connections.

- Connect the monitoring and control devices according to the specified wiring diagrams and guidelines.

TESTING AND COMMISSIONING THE SYSTEM

Testing and commissioning the solar energy system is a crucial step to ensure its proper functioning, safety, and compliance with regulations. These are the key steps involved in testing and commissioning:

1. Visual Inspection:

Conduct a visual inspection of the entire system, including solar panels, mounting structures, wiring, inverters, charge controllers, and other components. Check for any physical damage, loose connections, or abnormalities.

2. Electrical Testing:

- Measure the open-circuit voltage and short-circuit current of each solar panel string to ensure they are within the expected range.
- Measure the voltage and current at various points in the system, such as at the output of the solar panels, combiner box, and the input and output of the inverter or charge controller.
- Use a multi meter or other appropriate testing equipment to verify the continuity and integrity of wiring connections.

- Confirm that the grounding system is properly installed and functioning.

3. Performance Testing:

- Monitor and record the energy production of the solar energy system over a specific period, such as a few days or weeks, to assess its performance. Compare the actual energy production with the expected or estimated values.

- Analyze the performance data and address any discrepancies or issues that may arise. This could involve troubleshooting faulty components, optimizing system settings, or fine-tuning the positioning of the solar panels.

4. System Functionality:

Test the functionality of the monitoring and control devices, such as data loggers, energy meters, or smart home integration. Ensure that the system is providing the intended monitoring, data collection, or control capabilities.

5. Safety Checks:

- Verify the proper operation of safety features, such as ground fault protection, surge protection, and overcurrent protection devices.

- Ensure that all wiring and connections comply with local electrical codes and safety standards.

- Confirm that warning labels, signage, and safety instructions are appropriately placed and visible.

6. Grid Connection (for Grid-Tied Systems):

- If you have a grid-tied system, coordinate with the utility company to complete the grid connection process. This may involve inspections, meter installation, and paperwork to establish net metering or feed-in tariff agreements.

- Ensure that the system meets the necessary requirements and passes any required inspections or approvals from the utility company or relevant authorities.

7. Documentation and Handover:

- Prepare a comprehensive documentation package that includes all relevant information about the solar energy system, such as system specifications, component warranties, system design, wiring diagrams, test results, and compliance certificates.

- Provide the documentation package to the system owner, along with a thorough explanation of the system's operation, maintenance requirements, and any relevant warranties or guarantees.

www.ingramcontent.com/pod-product-compliance
Lightning Source LLC
Chambersburg PA
CBHW060849220526
45466CB00003B/1301

* 9 7 9 8 3 9 5 2 3 1 0 5 5 *